Is This Really My Body?

Embracing Physical Changes

ABDO
Publishing Company

Is This Really My Body?

Embracing Physical Changes

by Holly Saari

Content Consultant
Dr. Robyn J. A. Silverman
Child/Teen Development Expert and Success Coach
Powerful Words Character Development

Credits

Published by ABDO Publishing Company, 8000 West 78th Street, Edina, Minnesota 55439. Copyright © 2010 by Abdo Consulting Group, Inc. International copyrights reserved in all countries. No part of this book may be reproduced in any form without written permission from the publisher. The Essential Library™ is a trademark and logo of ABDO Publishing Company.

Printed in the United States.

Editor: Amy Van Zee
Copy Editor: Melissa Johnson
Interior Design and Production: Becky Daum
Cover Design: Becky Daum

Library of Congress Cataloging-in-Publication Data
Saari, Holly.
  Is this really my body? : embracing physical changes /
by Holly Saari ; content consultant, Robyn J.A. Silverman.
       p. cm. — (Essential health : strong, beautiful girls)
  Includes index.
  ISBN 978-1-60453-751-2
  1. Teenage girls—Growth. 2. Teenage girls—Physiology. 3. Teenage girls—Health and hygiene. I. Title.

RJ144.S23 2010
613'.0433—dc22
                                   2009002133

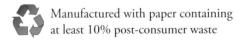

Contents

Meet Dr. Robyn

Dr. Robyn Silverman loves to spend time with young people. It's what she does best! As a child and adolescent development specialist, Dr. Robyn has devoted her time to helping girls just like you become all they can be. Throughout the Strong, Beautiful Girls series, you'll hear her expert advice as she offers wisdom on boyfriends, school, and everything in between.

An award-winning body image expert and the creator of the Powerful Words Character System, Dr. Robyn likes to look on the bright side of life. She knows how tough it is to be a young woman in today's world, and she's prepared with encouragement to help you embrace your beauty even when your "frenemies" tell you otherwise. Dr. Robyn struggled with her own body image while growing up, so she knows what you're going through.

Dr. Robyn has been told she has a rare talent—to help girls share their wildest dreams and biggest problems. Her compassion makes her a trusted friend to many girls, and she considers it a gift to be able to interact with the young people who she sees as the leaders of tomorrow. She even started a girls' group, the Sassy Sisterhood Girls Circle, to help young women pinpoint how media messages impact their lives and body confidence so they can get healthy and get happy.

As a speaker and a success coach, her powerful messages have reached thousands of people. Her expert advice has been featured in *Prevention* magazine, *Parents* magazine, and the *Washington Post*. She was even a guest editor for the Dove Self-Esteem Fund: Campaign for Real Beauty. But she has an online presence too, and her writing can be found through her blogs, www.DrRobynsBlog.com and www.BodyImageBlog.com, or through her Web site, www.DrRobynSilverman.com. Dr. Robyn also enjoys spending time with her family in Massachusetts.

Dr. Robyn believes that young people are assets to be developed, not problems to be fixed. She's out to help you become the best you can be. As she puts it, "I'm stepping up to the plate to highlight news stories gone wrong, girls gone right, and programs that help to support strengths instead of weaknesses. I'd be grateful if you'd join me."

Take It from Me

Change isn't always easy. It's even harder to deal with when there's a lot of it in a very short period of time. In the coming years, your body will develop and go through quite a few stages that will be new to you. Learning about what you will face is an important part of preparing for and dealing with these changes.

Having a positive and healthy attitude about your body during this time is helpful. But let's face it, this isn't always easy because every girl and every woman has insecurities about her body—whether it is her size, skin, odor, or hair. Even models have stated they aren't as comfortable with some parts of their body as others—and they are retouched to look perfect! So how are you supposed to feel comfortable being yourself when the entire physical part of you is busy changing?

My hope in writing this book is to help you get comfortable with yourself by learning the changes and stages your body will encounter. You'll read other girls' stories and hopefully realize you're not the only one out there going through these things. It's reassuring to find that many girls before you have experienced the very same situations.

Having braces and bad hair for a while is natural. By no means is it the end of the world, although at times it sure can feel that way. We have all walked that path. Just remember, you are not alone.

XOXO,
Holly

1

A Normal Body

*L*ook around you. Does everyone you see have the same body type as you? Or do you see a wide variety? It's no surprise that bodies come in all shapes, sizes, and types. Think of how boring it would be if we all looked exactly the same!

Yet this differentiation usually doesn't happen to girls' bodies until adolescence. Until then, lots of girls' bodies can look pretty similar. During adolescence, girls' bodies grow and change. Girls who used to be slender may develop hips and girls who were more filled out may grow into their curves.

The trouble is, this growth period can be a bit overwhelming. It's hard to deal with your body changing in ways you've read about but that are still new to you. It's even harder to deal with your changing body when it develops in ways you're not comfortable with or that the media says aren't natural or beautiful. (But they are!)

How does a girl deal with these body changes, especially if the results are different from how she thinks she should look? How does a girl deal with the body part that she just can't stand? Melinda had this very question, and with some help, she began to find an answer.

It's hard to deal with your body changing in ways you've read about but that are still new to you.

Melinda's Story

Melinda had always been relatively thin. But the past summer had been a changing point for Melinda. Her body had started developing more. In truth, she had gained a few pounds—and all in her love handles, she thought. When she was shopping for new school clothes for the fall, Melinda turned down everything because she felt her midsection looked so awful in each outfit she tried on. Her mom would tell her that she looked great but that just made things worse. Melinda knew her mom had to say those things because she was her mom. They weren't true.

Talk About It

- Do you have trouble finding clothes that fit your body? What do you do about it?

- Do you believe your friends and family when they give you compliments? Why or why not?

Starting seventh grade was stressful enough without thinking she'd be the chunkiest girl in her class. The night before her first day, Melinda laid out her clothes on her floor. They weren't new, but they fit. Throughout the summer, the old jeans she planned to wear on her first day had stretched out as she'd grown. No other pants she had tried on while shopping fit her in the midsection and the thighs. So, she would just have to wear the old, stretched-out pair. Melinda hoped no one would notice her clothes from last year.

Talk About It

- Have you ever wanted to wear a new outfit but nothing you tried on fit? How did you feel? What did you do?

- Do you have a body part that you're uncomfortable with, like Melinda? What do you do about it?

- How do you prepare for the first day of school, or for an important day or event? Do you plan in advance?

But people did notice—not that her pants were old, necessarily. Some girls noticed how she was wearing them. Over the summer, Melinda had started hiking her pants up really high so the waistband would cover her middle and not make her sides bulge over

the top of her pants. She didn't think anything of it, but some of her classmates did. It was worst in her biology class when two popular girls in her grade said something to her about it.

"Hey, Melinda," Erin said. "Nice pants. If you pulled those up any higher you wouldn't have to wear a bra." Sera giggled next to her at the lab table.

"Yeah, Melinda," Sera said. "Your butt looks as tall as a billboard." The two girls kept laughing while Melinda sat mortified. They had spoken so loudly that Melinda was sure the whole class must have heard. She could feel the tears coming to her eyes. She mumbled something to her teacher about feeling sick as she left the classroom and didn't look back.

Talk About It

- **How would you feel if you were Melinda? What do you think about Melinda's reaction?**

- **Do you know girls who tease other girls in school? Have you ever teased anyone? If so, why?**

- **Has anyone ever teased you about how you looked? How did you feel? How did you react?**

Melinda ran to her locker and grabbed a sweatshirt. She tied it around her waist so that it covered her backside. She hoped it would be enough to keep anyone else from making fun of how she looked.

During the rest of the day at school, Melinda was totally preoccupied with what had happened in biology class. She couldn't believe that Erin and Sera had said such mean things, and she imagined that everyone was talking about it. She didn't know how she would be able to go back to school the next day and face her classmates.

When Melinda got home from school, it was obvious she'd been crying. Her older sister, Amy, was sitting at the kitchen table.

"What's wrong?" Amy asked.

Melinda had planned to keep what happened a secret, but she started crying and ended up telling her sister the whole story.

"Well, Melinda, I see what they're saying—sort of. Obviously they were mean in how they said it, but you're trying to wear pants that aren't really good for you anymore. Come on, I'll help you find something for tomorrow."

And although Melinda still had tears in her eyes, she trekked up the stairs after her sister. After trying on a few outfits that Amy picked out for her, Melinda found something that fit her nicely—and that she felt comfortable wearing. While she still felt the sting of her classmates' comments, Melinda hoped that the next day at school would be much better.

Talk About It

- Do you think Amy handled the situation in a good way? Why or why not?

- Have you ever confided in a family member, relative, or close friend when something has hurt your feelings? What happened afterward?

- Do you have someone who could help you with situations like Melinda's?

Girls today are bombarded with unhealthy images of women. These images are supposed to show what actual women's bodies look like, but instead show examples of unattainable bodies. Did you know that the models in the advertisements don't even look like that in real life? They're retouched to make some body parts smaller and others bigger!

When we compare ourselves to model-like images, of course we find ourselves lacking. We're comparing ourselves to someone who is retouched to look "perfect." Even so, many girls may not only find their flaws, but obsess over them and blow them out of proportion. Beautiful girls like Melinda may think they're abnormal because they have curves that they don't see in magazines.

It's important to realize that your body will continue to change for some time after adolescence. As long as you have healthy habits, your body will adjust and come into its own soon enough. If, during that middle period, you feel a bit uncomfortable or awkward about your body, there are tips that can help you. Just remember, there is no one body type. Understanding yours can be a key to becoming comfortable and confident in your own skin.

Get Healthy

1. Wearing clothes that flatter your body can make you feel better about yourself. Look through your wardrobe to see what you can mix and match to complement your best attributes and tone down the ones you aren't so comfortable with.

2. It can be refreshing to share insecurities and have a good laugh about them. Share with a good friend how you feel about your body and ask her how she feels about hers. You both may realize that you make more out of your insecurity than you should.

3. Try to decrease how often you view unrealistic female images from magazines, television, and other media sources. Look for images of real female role models instead of comparing your body to retouched, unhealthy models.

The Last Word from Holly

Finding clothes that fit and flatter is not always an easy task. In fact, it's often quite a challenge. But if you think of it as an exciting challenge, you are already on your way to looking and feeling your best. Remember, we all have parts of our bodies that we're not in love with. It's all about accentuating the positive and finding ways to feel good about yourself and your body as a whole.

2

Breaking Out

You wake up one morning and look in the mirror and there it is, staring right back at you: a giant zit. Nobody likes to have acne. However, acne is probably the greatest common denominator in adolescence. Very few people, both guys and girls, grow up without having occasional breakouts.

Having acne may make those everyday tasks, such as talking to a friend or speaking in class, quite unpleasant. Acne can dramatically affect self-esteem and confidence. Girls with acne may be so embarrassed, or even ashamed, about how they look that they may withdraw from others and their favorite activities because they don't want anyone to see them.

Luckily, there are many medical treatments for acne, both in drugstores and through a dermatologist. However, a girl needs to do her part by having good hygiene practices as well. Hannah has started to develop acne. Read her story to see if she deals with it in a healthy manner.

Hannah's Story

Hannah was close to her older sister, Crystal. Crystal was 15, so she had already gone through most of puberty, including getting acne break-

Having acne may make those everyday tasks, such as talking to a friend or speaking in class, quite unpleasant.

outs. When Crystal got a zit that was ready to squeeze, she would call Hannah to come into the bathroom to watch as Crystal popped it. It was sort of an odd tradition that the sisters shared.

Hannah and Crystal's mom knew about this activity and would scold the girls.

"That is not going to help a thing. I'm warning you right now that it'll only make things worse," she would say.

Crystal and Hannah would laugh off their mom's comments, which didn't seem true at all. Crystal only got pimples on very rare occasions. Usually, on the morning after Crystal squeezed one of her zits, the blemish had nearly vanished. If it was still a little red, Crystal would deftly apply some of her cover-up, and

her face would look spotless. She'd be ready for school with no one knowing about the blemish but herself and Hannah.

Talk About It

- Have you started to develop acne? If you have, how do you feel about it?

- Have you ever tried squeezing a zit? What happened afterward? Did it go away or did it get worse?

- Have you ever gotten advice from a parent or other adult that you ignored? What happened in the end?

Soon, Hannah began to feel a couple of bumps on her face. A few days later, she woke up and saw them: two pimples on her chin and one on the side of her nose. Hannah was mortified—she wanted to hide her face until the end of time, but she knew she had to go to school.

"Crystal!" Hannah screamed down the hall. "I have these three huge zits, and I don't know how to fix them. You've gotta help me!"

As Hannah stared at her face in the bathroom mirror, Crystal meandered in.

"Let me see," she said, grabbing Hannah's chin to inspect her face. "You are so melodramatic, Hannah. These aren't that bad. I'll cover them up with some makeup, and you'll be fine for school." Crystal grabbed her makeup kit and worked a little magic. Hannah examined her craftsmanship and approved. She'd be okay for the school day.

Talk About It

- Have you ever woken up one morning and saw you had a large pimple when you least expected it? What did you do?

- Do you use makeup? How often do you wear it? Do you use it to cover up blemishes or for other reasons?

At dinner that night, Hannah was antsy to leave
the table and look in the mirror. She had felt her
pimples periodically throughout school, and she felt
them again at the table. She thought they were ready
to pop, because they hurt a lot. As soon as Hannah
cleared her plate she raced to the bathroom and called
Crystal in too.

Hannah tried to squeeze her pimples. It was her first time doing so, and Crystal had to instruct Hannah on how to do it. Hannah tried squeezing one on her chin first, but nothing happened. She pressed and pressed, but nothing happened—except after a while, her chin throbbed and her fingernails were imprinted in her skin.

"Crystal, why didn't that work?" Hannah asked.

"Maybe it's just not ready. Or maybe you didn't do it right. I don't know. Try the one on your nose," Crystal said through laughs.

Hannah rolled her eyes and tried, but nothing happened. Hannah was annoyed. She still had these zits, and her face hurt even more now. She was so annoyed that she went to bed right away, and forgot to wash her face before going to sleep.

Hannah tried squeezing one on her chin first, but nothing happened.

Talk About It

- **Do you think the activity that Hannah and Crystal share is healthy?**
- **Do you wash your face every night?**

When Hannah woke up the next morning, she had already forgotten about the previous night's events. But when she shuffled into the bathroom and looked at her face, she gasped in horror. All along her chin there were zits. She quickly counted. There had to be at least ten. Some were small, some were larger, but they were all red and looked like a little grove of chicken pox. How could this have happened?

"Crystal! Mom!" she called. "Anyone!" Both appeared at the bathroom door. The new pimples were like a horrible case of déjà vu, but now it was so much worse. There was no way makeup would cover them all. And there was no way she was going to school looking like this. When Crystal saw Hannah's chin, her mouth opened, and then she sort of laughed.

"Don't laugh, Crystal, this is your fault! Mom, I can't go to school looking like this. Can I please, please, please stay home today?" Hannah pleaded.

Talk About It

- **What do you think Hannah's mom will say? Have you ever wanted to stay home from school for a reason other than actually being sick? Did you get to?**

- **What do you think Hannah has learned about picking at her pimples and not washing her face?**

Acne is a large concern among adolescents. So, what can be done to prevent or lessen the severity of acne? Knowing the causes of acne, and knowing what is in your control, can be a start. Two large factors behind acne are hormones and hygiene. While you cannot really control the hormonal fluctuations of adolescence, you can control your personal hygiene. This means not picking at your pimples. Doing so only makes things worse, as Hannah found out. And believe it or not, picking at your face and squeezing zits can become pretty addictive. Worse, it can lead to scarring.

Another cause of acne can be stress. Isn't it odd that at times when we are stressed out and least want a pimple to arrive, we spot one in the mirror? Keeping your stress in check and including activities in your life that help you cope with stress are great tools to decrease the amount of stress-related acne you have.

Get Healthy

1. Keep your face clean. Wash it each morning and night with an oil-free soap. Don't scrub too hard, and use a moisturizer. If you wear makeup, be sure not to wear it overnight. It may settle into your pores, eventually clogging them.

2. Drink a lot of water. Your skin needs hydration and drinking water helps your body flush out the toxins that could cause breakouts.

3. Do not pick, pop, or squeeze your acne. Although it may seem necessary or natural, doing these things will only make it worse. The zit could become inflamed, infected, and even larger and worse than it would have if you left it alone. Plus, it could leave a scar.

4. If you have constant breakouts of many pimples, consider seeing a dermatologist. Your doctor may prescribe some medication to control your acne.

The Last Word from Holly

Acne is the last thing a girl wants to have. Not all acne can be prevented all the time. Unfortunately, it's a part of adolescence that isn't very cool and can be downright painful, both physically and emotionally. That said, there are things you can do to lessen the amount of acne you have. Taking the steps to have good hygiene, such as washing your face and drinking lots of water, can help. I hope you can realize you are so much more than skin deep. But I do understand that it's difficult to see, especially when what you see in the mirror is acne. Just hang in there.

3

Unique Patches

While not as common as acne, birthmarks can be a big source of embarrassment. Unlike acne, many birthmarks are not easy to cover up. This can make it frustrating for some girls who have birthmarks in more noticeable places—on their face, neck, or arms. Birthmarks usually form from extra pigment in the skin or blood vessels that have been clustered together. They come in different colors, shapes, and textures and can appear on different places of the body.

Having a birthmark or another skin condition does not make you a freak—on the contrary, it makes you unique. Being different from the rest of your peers may be the last thing you want to be at this point in your life, but having interesting characteristics gets cooler as you grow older. For the time being, if you are self-conscious about a birthmark, there are some tricks to make them less conspicuous, as Josie finds out. Read her story to find how she deals with her birthmark.

Josie's Story

All her life, Josie had a reddish-purple birthmark on her shoulder and collarbone. She hated it, and whenever she would catch a glimpse of it in the mirror, she couldn't help but feel abnormal.

Luckily, Josie's birthmark was on a place of her body that was usually covered by her shirts. In fact, Josie only purchased tops that would cover her birthmark. Not many people outside her family knew that she had it, because she was careful to hide it at all times. Josie never went to the beach with her friends, and she never wore tank tops in the summer because her birthmark made her insecure.

Josie never went to the beach with her friends, and she never wore tank tops in the summer because her birthmark made her insecure.

Although at most times she could cover her birthmark, Josie still was horrified that she had it. She often looked through magazines and wished to have regular skin like all the models. Whenever she would see other girls in tank tops or spaghetti-strap dresses, she felt a surge of jealousy and self-loathing. She saw her birthmark as a disfigurement, not as something that made her unique.

Talk About It

- Do you think Josie is overreacting? Why or why not?

- Do you have any body parts or aspects of your body that you don't like at all? What are some things you can do to come to terms with it?

- Picture yourself in Josie's position. What would you do? How would you feel?

Covering her birthmark had been an easy thing for Josie in the past. This year, however, she had her first formal school dance coming up in a few weeks. Josie had a crush on a boy in her class named Kevin. So, she was completely thrilled when he asked her to go to the dance with him. Josie couldn't believe her luck. She and her friends couldn't stop talking about how great the dance would be. They also talked about the dresses they wanted to wear.

Josie was so excited that her crush had asked her out that she totally forgot about having to wear a fancy dress. Her birthmark was sure to show then! Josie had no idea what she was going to do.

Josie talked to her mom about the dilemma. Her mom knew how embarrassed Josie was about the birthmark. Her mom mentioned something that the two had talked about before.

"Well," her mom said, "I think it's time we tried some makeup on it."

"Do you think it'll work, Mom? Because if it doesn't, I am not going to the dance. Kevin can't see how much of a freak I am," Josie lamented.

"Josie, you need to stop thinking you're a freak because you have a birthmark. I don't want to hear you calling yourself that anymore. And I have a good feeling about this makeup. We'll pick some up this weekend when we look for a dress."

"Okay," Josie said. That's all she could say before she started squealing with excitement. She just hoped that the makeup would work.

Talk About It

- **How does Josie's relationship with her mom help her?**
- **Are you close with your parents or guardians? Do they know your insecurities about your body and try to help you with them?**
- **What do you think will happen with Josie, her dress, the makeup, and the dance?**

Josie stood in front of her bedroom mirror in her formal dress. She felt like a movie star in the blue, floor-length gown. Her mom had curled her hair and

pinned it up, and she was wearing some mascara and lip gloss.

The only thing Josie had left to do was cover her birthmark. She was really nervous. Josie had tried the makeup as a test run, and she found that it didn't work as well as she had wanted it to. She was hoping this time it would cover the birthmark completely. Josie asked her mom for some help. Maybe she would have some tricks up her sleeve.

When her mom was done applying the makeup, Josie looked in the mirror and was a little disappointed. She could still see her birthmark. Yes, it was much lighter, but Kevin and her friends would notice it, she was sure.

Josie got one more pep talk from her mom before the doorbell rang. This was it, Josie thought. No turning back now.

Josie got one more pep talk from her mom before the doorbell rang. This was it, Josie thought. No turning back now. Her mom opened the door, and her date walked in.

"Hi, Kevin," she said.

Talk About It

- **What do you think Josie was thinking right before she saw Kevin?**

- **Do you think Josie was thankful for her mom's help that night? Have you ever dressed up for a formal dance? How did it make you feel?**

- **What do you think will happen at the dance? Do you think Kevin or Josie's other friends will notice her birthmark?**

Ask Dr. Robyn

During adolescence, fitting in is a big deal. Having anything that makes a person stick out can be a dreaded thing. And because birthmarks don't happen to the majority of girls, having one—or another skin condition, for that matter—can make a girl feel really insecure.

Many people accept their birthmarks as a distinctive, cool, or interesting part of their bodies. These people see their birthmarks as unique—something that sets them apart from the crowd. But if making yours less noticeable is what you really want, there is probably a way to do that. There might be an over-the-counter method, or you may need a more expensive medical procedure.

While a girl's biggest concern about her birthmark may be appearance-based, she should also consider health concerns as well. Although most birthmarks are harmless, some can suggest underlying health problems or lead to them in the future. One kind of birthmark can increase one's chances of developing skin cancer in later years. See a doctor if you do have a birthmark. You can probably learn if there is any need for treatments or any potential health risks in one visit.

Get Healthy

1. If you have a birthmark, do some research to see what kind of birthmark it is and if there may be an underlying medical issue. Whether you feel there may be reason for concern or not, it's not a bad idea to get your birthmark checked out by a doctor.

2. Keep in mind that birthmarks can be beautiful. After all, people get tattoos on purpose. Some people are born with these distinctive features.

3. Makeup today is pretty amazing. Birthmarks and other skin discolorations can often be covered up. Several brands of makeup are designed to camouflage birthmarks. If covering up your birthmark is important to you, consider investing in a good concealing makeup.

The Last Word from Holly

It is common to have a body imperfection that you are self-conscious about. Retouched images in the media do not help the situation. But it helps to remember that people aren't going to be focused on our negative aspects just because we are! With time you'll see that the more you accept the things that make you who you are, the more others will see your imperfections as the things that make you beautiful.

4

Bad Hair Day

Hair is changeable, making it one of the easiest ways to alter our appearance. Yet, when something goes wrong—a haircut or a style—it's difficult for us to see past what our hair looks like at that moment. It's sometimes hard to realize that hair will grow out and there's a fresh start coming. It may even be hard to consider that hair can be covered up in lots of different ways.

Hair is something that may change during puberty. Doctors think that hormones affect the texture of a person's hair. So that means that while you may have had curls as a child, you may enter into your high school years with stick-straight hair. These hormones may also make your hair greasy or dry.

Let's face it, hair is a big deal. And when it's not how we want it to be, we can get upset. Not having the texture, style, or length of hair we want is a problem, even if we may blow it out of proportion. It's still frustrating. And what girl hasn't had a bad haircut that put her on the verge of tears? Our hair can show others who we are as people. When it's messed up, we can feel messed up too.

> **Let's face it, hair is a big deal. And when it's not how we want it to be, we can get upset.**

Take Rachel, for example. When she gets an overly ambitious haircut, she experiences why hair issues can be very emotional.

Rachel's Story

"I don't know, Rachel. I just don't think that cut is a good idea."

Rachel was annoyed with her stylist, Leslie. She had been coming to her for a year now, and Leslie had always been great. But now, she kept telling Rachel that the cut she wanted would not work for her.

"No, I think it will work," Rachel tried to reason, "because I'm going to start straightening my hair every morning. So it'll look exactly like this."

"Rachel, that's going to be a lot of work," Leslie said. "You've always told me you like a low-key cut so you can wash and go. I didn't think a lot of styling was

for you. Why don't you think about it for a few more minutes while I prep my station?"

Rachel was annoyed. She looked at herself in Leslie's mirror. Even though she had straight hair when she was younger, her hair had recently become really difficult to manage. It was super frizzy and wavy and reached the middle of her back. She looked down at the magazine photo she held showing a girl with a straight bob that reached a little higher than her shoulders. It was so cute, Rachel thought. She used to have straight hair, so she could totally look like that. She saved her money for the flat iron she had just bought. Now all she needed was the haircut.

Leslie looked at her with questioning eyes. "So? What's your final answer?" she asked.

"Let's do it," Rachel said.

Talk About It

- Do you think the new haircut Rachel is getting is a good idea? Why or why not?

- How do you feel about your hair? Do you have a tough time working with its texture and type? Do you ever want a completely opposite look, like Rachel?

A little while later, Leslie turned Rachel around in the salon chair and Rachel saw her new style. She couldn't help but beam. She loved it! Leslie had straightened Rachel's hair so that it looked nearly identical to the magazine photo. Sure, it had taken nearly an hour of blow drying and flat ironing to get it that straight, but it was worth it. Rachel couldn't wait to see her friends' reactions the next day at school.

But the next morning, Rachel got up at her usual time, forgetting that she needed at least another hour for her new hairstyle. And her hair was too greasy to go without a washing. Rachel hopped in the shower and got ready in her usual whirlwind way. Her

Rachel hopped in the shower and got ready in her usual whirlwind way.

hair had dried a bit by the time she was eating breakfast, and she caught a look at herself in the hallway mirror. Oh no, she thought, her hair was starting to frizz, and it was even more wavy than it had been when her hair was long and weighed down. There was no way she could put her hair in a ponytail now because it was too short. She had to get out the door for school, so she had to think fast. Rachel grabbed an old baseball hat from the hall closet, shoved it on her head, grabbed her stuff, and flew out the door.

By the time she reached school, her hair had completely dried, and Rachel had felt her hair expand underneath her cap. There was no way she could take her hat off or she would have the biggest poof ball.

"Ms. Connelly, you know the school policy," her science teacher said as Rachel slipped into her seat in first period. "Please be so kind as to remove your hat, so we can all begin our education."

"Umm," Rachel tried to stall. She completely forgot they couldn't wear hats in school. She had no idea what to do.

Talk About It

- When it comes to your hair, are you a wash-and-go girl, like Rachel? Or do you leave enough time to style your hair?

- What do you think Rachel is thinking and feeling right now?

- Have you ever inadvertently forgotten a school rule before? What happened?

Rachel looked like a deer in the headlights. Her options were to refuse to take off her hat and probably get sent to the principal's office, or take off her hat and be absolutely mortified that her classmates would see her frizzy hair. Rachel slowly took off her hat, while fervently pressing her hair down, silently praying that the hat had helped smooth it.

As soon as the hat was off her head, a couple of boys started giggling.

"What happened to your head, Rachel?" one of them asked. "Did you run into a chain saw over the weekend?" The two boys laughed even harder, and were joined by more of her classmates. It seemed everyone was looking at her. Two girls were whispering and staring at her with horrified looks.

"Umm," Rachel said. She had no idea how to respond.

"All right, people, show's over," her teacher said. "If it's all right with you, I'd like to start teaching now." And with that, her teacher started his lesson. But nearly every other minute, Rachel could feel someone sneaking a glance at her hair. Rachel slouched down as low as she could in her chair and used all her might to hold back her tears. How would she deal with this all day? And every day in the future?

Talk About It

- What would you do in Rachel's situation? Would you take off your hat? Or would you refuse and risk getting in trouble?

- Have you ever been embarrassed in front of your classmates? What happened? How did you deal with it?

- What advice would you give Rachel?

Hair issues can be traumatic, and learning how to style your hair can be a challenge. Hair styling takes practice and patience. This is good to remember on those bad hair days when nothing seems to be going right for you. With time, you'll learn the skills you need to fix your hair.

This flows directly into another hair concern among adolescent girls. Most are rarely satisfied with their own hair type and wish they had a completely different one: those with curls want straight hair and vice versa. But getting used to your hair type and accepting your hair for all its beauty and uniqueness is a good goal to have.

Bad hair days are going to happen. Bad haircuts are going to happen. They have happened to us all. But chances are the next day won't be such an awkward hair day. And don't forget that bad haircuts will grow out, giving you the opportunity to start fresh.

Get Healthy

. Work with your hair type. If you have naturally curly hair, perhaps that short, sleek bob you had your heart set on will not turn out as you envisioned. If you really want a change, ask your stylist what kind of cut and hairstyle would work for your hair type

2. Go in baby steps with new cuts or styles to avoid anything drastic that you may regret. It will also help you get more used to your big decision in the end.

3. If you want some change, add variety to your hairstyles. Wear a ponytail one day, all your hair down the next, and your hair half up after that. Maybe you can try curling or straightening your hair. Use headbands, clips, or other accessories for added variety and style. Again, ask your stylist for some tips!

The Last Word from Holly

Hair is a big deal—it's a big part of a girl's overall look! But just like zits or a birthmark, hair is not all there is to a person. While a cute or stylish haircut can make a real statement, don't forget that it's only a part of who you are. If things go awry and you leave a salon in tears, remember that just because your hair looks different doesn't mean you've changed. A bad haircut is not the worst thing in the world. With a little time, your hair will grow. And in the meantime, get creative with headbands, pins, or scarves. Make your hairstyle work for you!

5

Brace Face

Braces have a bad rap. Who hasn't heard the story about a boy and a girl with braces who kissed and got stuck together? With this sort of folklore attached to them, braces hardly seem a glamorous beauty tool. But in truth, braces are a brilliant invention. They transform crooked teeth into wonderfully straight smiles.

But along with the stories of braces, there is real paranoia and embarrassment when you have them on your teeth. You may fear your mouth looks full of metal or that you'll smile with food in your teeth. You might be nervous about being the only one with braces. In an extreme case, this paranoia and insecurity may affect a girl's confidence to the point of altering her regular habits, and even her moods. This was the case with Mia. Read her story to see how braces started affecting her actions.

Mia's Story

Mia was the first of her friends to get braces. She was in fifth grade, but her orthodontist said her teeth and mouth were ready. She had a large gap between her top two teeth, and two of her other teeth pointed forward. Her orthodontist said it might take awhile for her teeth to straighten out, so it would be best to start as soon as possible.

Although there were other options, Mia got traditional metal braces on her top and bottom teeth. The first few days after her braces were on, Mia's teeth and entire mouth were really sore. She could barely chew and had to eat foods that were mostly liquid, like soups and breakfast shakes. Mia found it hurt less if she moved her mouth as little as possible. So she only talked when it was necessary, using as few words as she could. Smiling even hurt, so she tried to stop smiling as much as well.

Her orthodontist said it might take awhile for her teeth to straighten out, so it would be best to start as soon as possible.

Talk About It

- Do you have braces? When did you get them? How long do you have to wear them?

- Has your mouth ever been so sore that you could barely eat or talk? How did this affect you?

Mia's mouth felt nearly back to normal by the end of her second week with braces. Mia could now eat solid foods and her mouth didn't ache when she talked or cracked a smile. Yet, when Mia caught a glimpse of herself smiling in a mirror, she was shocked. It was the first real look she had gotten at her metal mouth, and she thought she looked ridiculous—like a weird robot or something. And because none of her friends had braces yet, Mia felt even more self-conscious. Maybe people wouldn't be as grossed out by her if she didn't let them see her braces that much. Mia decided she would try to keep her mouth closed as much as possible.

Over the next few days, Mia worked at controlling her emotions. When she felt like smiling, she didn't. When she felt a laugh coming out, she tried to suppress it. Mia rarely spoke up in class anymore, and she was quieter among her friends.

To Mia, it was all worth it. She would rather be quiet and have nobody notice she was there than to have everyone stare at her braces.

Talk About It

- **Do you have something that you're embarrassed about, like Mia is? How do you deal with this?**

- **What would you do in Mia's situation?**

- **How do you think Mia will be affected by not smiling or laughing as much? Do you think anyone will notice?**

Mia was surprised at how easy it was to stop showing her teeth and braces. But as the days passed, and Mia continued not to smile or laugh, her mood and

attitude began to change. Mia was sadder. Although she usually had boundless energy, she became lethargic. And while Mia became used to being quieter, she didn't express her feelings to her friends like she always had so easily.

One day, Mia's friend, Ashley, sat by her at lunch. "Mia, is everything okay? You've been really quiet lately," Ashley asked.

"I'm fine," replied Mia. "Why do you think something's wrong?"

"Well, I don't know, Mia. You seem a little down. I'm not sure really, but I haven't seen you smile or laugh in days. It's just weird. I want to make sure everything's okay, I guess," Ashley said.

"I'm trying not to show my ugly braces as much. So that means I can't laugh or smile like I used to."

"No, that's my plan, Ashley," Mia said. "I'm trying not to show my ugly braces as much. So that means I can't laugh or smile like I used to."

Ashley's mouth gaped open. "Mia," she said, "that's ridiculous. How can you just stop smiling or laughing? And, have you seen all our teeth? We'll all get braces sooner or later, and I for one refuse to stop having fun because of that. Come on, you have to lead the way!"

Talk About It

- **What do you think of Ashley's response to Mia's plan?**
- **What would you have said if Mia was your friend?**
- **How do you think Mia will act now?**

After talking with Ashley, Mia began to think about how not smiling had been affecting her mood. In truth, she really had been feeling lonely and sad by isolating herself so that nobody would notice her braces. She missed laughing with her friends, and she decided to do something about it.

That night, Mia called Ashley on the phone.

"Mia! Hi! What's up?" Ashley asked.

"Well, I thought about what you and I talked about at lunch today, and I agree with you," Mia said. "I'm sick of not smiling. What are you doing Friday? I was wondering if you and Josie wanted to sleep over. I thought we could rent that new funny movie about the dog sitter. I could use a good laugh."

Talk About It

- Why do you think Ashley's comments had such an effect on Mia?

- Do you think it is a good idea to tell your friends when you are concerned about them? Why or why not?

- What do you think of Mia's new plan? Do you think she's doing the right thing?

According to the American Association of Orthodontics, the average age for braces is between 9 and 14 years old and the average time span a person has braces is two years. These statistics show that braces will most likely come at a time when the angst, confusion, and drama of adolescence are at a high point. But there is good news about braces. First, braces have changed a lot since I was a teenager. Braces don't have to be made of metal. They can be made of clear plastic, making them a lot less visible.

Second, and most important, braces make your teeth straight. So while the period that you have them on can be uncomfortable and embarrassing at times, your braces will come off to reveal a beautiful smile. The time you have your braces on will be overshadowed by all the time you will have wonderfully straight teeth. And don't forget that you won't be alone—chances are many of your friends and classmates will have them too. So like Mia realized, it's not worth it to change your personality because you have braces. It's okay to look forward to getting them off, but in the meantime, remember that you are beautiful, braces and all.

Get Healthy

1. Keep good dental hygiene while you have braces. There would be nothing worse than to get your braces off to find plaque-covered teeth and swollen gums. Ask your dentist about special tools for flossing with braces.

2. Have a handheld mirror in your locker to check your teeth after you eat. Double-checking that your teeth are food-free will allow you to smile with confidence.

3. Don't let having braces change who you are. When you're confident—smiling and laughing—others will see that you are so much more than what is on your teeth.

The Last Word from Holly

I had braces in seventh and eighth grades, but it really wasn't too bad. Most of my girlfriends had braces around the same time I did. Around the lunch table, we would show each other our teeth to make sure bits of lunch weren't hiding out in our dental work. Don't get me wrong—the mortification of having food stuck in our teeth was still overwhelming at times. But with my friends' help, I was able to get past the paranoia. Just remember, your braces will be off soon, and when you see your beautiful, straight teeth, you'll realize it was definitely worth it!

6

A Distinct Scent

Sweating is gross, isn't it? There's a good reason for it, though. Sweating is a way the body cools itself down and remains at a healthy internal temperature. Still, it isn't the most glamorous of bodily processes. And some girls may find it downright disgusting, especially if they sweat more than the other girls around them.

Not only does sweating make you feel unclean sometimes, but it can also lead to body odor. When sweat is released, it comes into contact with bacteria on your skin. This reaction causes the smell. Body odor usually begins around puberty because a hormone that contributes to the odor is released then. So body odor can hit at the time when you may be the most sensitive about your changing body. But don't be discouraged! There are ways to tone down, if not completely get rid of, body odor.

Knowing how to do this may not be second nature. Take Emily. She was a good student who took school seriously and did not cut classes—that is, until she felt cutting class was necessary for her to keep her humiliation in check.

Emily's Story

Emily liked school. She was in seventh grade and had always been a good student. She paid attention in class, handed in all her homework on time, and studied hard for tests. Emily was also a good athlete, and had played in a soccer club each spring and fall for the past three years. She loved playing the game and being on a team, but there was one thing about physical activity that she hated: Emily would sweat a lot. During soccer, it wasn't a huge deal because she was working hard and playing a sport—acceptable times to be working up a sweat.

But even the smallest amount of effort had her armpits wet and trickles of sweat gliding down her back.

But even the smallest amount of effort had her armpits wet and trickles of sweat gliding down her back.

Emily's excessive sweating had caused her a ton of anxiety, especially in school. Her locker was on the third floor, and each morning, and often between classes, she had to climb the flights of stairs to get her books. After the climb, she was sweating and continued to sweat until her heart rate calmed down.

Talk About It

- **Do you sweat when climbing stairs or doing other minor exertions? If so, how does it make you feel?**

- **What can Emily do to feel better about her sweating problem?**

Emily had noticed that whenever she would sweat, she started to smell. She put on deodorant every morning, and sometimes again during the day, but it didn't really help. She still had a body odor that was repulsive to her, and she didn't know what to do about it. She worried that others would notice.

Emily had phys ed during fourth period. Although she liked most of her classes, she hated phys ed because of her sweating and smelling. After phys ed class, Emily had three more classes before she was done with school. She was mortified that she had nearly half the school day left where her classmates could smell her body odor.

Unfortunately, Emily's phys ed teacher didn't give enough time to shower and get ready before the next class. Most of the other girls in Emily's class didn't even break a sweat, so they didn't need extra time to shower. For a while, Emily slipped out of the gym before the class was dismissed so she could get to the locker room.

She would wash her armpits and wipe off her sweat before the other girls could see and make fun of her.

Talk About It

- Do you sweat a lot during phys ed class? How do you feel after class? Do you have time to shower and freshen up before your next classes?

- What do you think Emily will do about sweating in phys ed class?

Lately, the phys ed activities had been more of a workout, and Emily realized that to feel completely clean, she needed to take a full shower. She knew she didn't have time. So, one day, right before heading to phys ed, Emily decided to skip class.

Emily had never cut a class before, but she didn't know what else to do. She was so embarrassed about her smell that she saw no other alternative. Emily changed into her workout clothes and walked into the gym. She stayed through the first part of attendance and instructions on the day's activity. Then she quietly snuck out and hid in the locker room for the rest of the period.

Emily had never cut a class before, but she didn't know what else to do.

Emily had a bad feeling about what she was doing. She was so nervous she'd be caught or that her phys ed teacher would notice her missing, but she felt certain that she had no other choice.

Talk About It

- Have you ever skipped a class? Why? How did you feel about it?

- What do you think will happen with Emily skipping phys ed?

- What are some other things Emily could have done?

Sweating and body odor are not attributes of the body that adolescent girls enjoy. They're not necessarily attributes adults enjoy either! But sweating happens, which is a very good thing, as it keeps the body's internal temperatures in equilibrium. In Emily's case, rather than skipping class, a quick talk with her phys ed teacher may have helped the problem. If her teacher knew how Emily felt, perhaps she could have allowed more time before the end of the period for everyone to freshen up.

As females, we may feel overly pressured to be clean, dry, and to smell fresh. While it may seem as if society says it's okay for boys and men to sweat during exercise or manual labor, females don't seem to get the same permission. This may also contribute to why you may feel more disgusted by sweating than boys your age do.

But try to remember that sweating is natural and healthy—for both genders. Without it, our bodies would overheat and there would be serious physical consequences. And remember, everyone does it! It's unfortunate that it can make us feel unclean, but you can take steps to keep yourself healthy and body-odor-free after considerable sweating. That way, you'll always feel your best.

Get Healthy

1. Just because you might sweat a lot doesn't mean you have to smell. If your deodorant isn't working, try a stronger, clinical-strength deodorant. You could even talk to your doctor about your concerns and see what he or she recommends.

2. If you need more time to shower after a gym class, talk to your teacher about it. It's a requirement in many schools for students to have the time to shower if they want it.

3. Keep good hygiene to prevent body odor. Shower every day if you tend to sweat heavily. This will keep body odor at a minimum and keep you feeling fresh and clean.

The Last World from Holly

Everyone sweats, and sometimes smelling goes with it. If you feel self-conscious about sweating, keep deodorant in your backpack or locker. When you're feeling uncomfortable about how you might smell, reapply some deodorant. Most girls don't like to sweat or smell because they feel gross, disgusting, or maybe even masculine. The best thing to do is be prepared for it when you know you'll be sweating, and to bathe regularly to keep your body clean and wash smells away.

7

Curvy or Not

As you get older, breasts are one of the most visual signs of your body's development. Media images bombard us with ways breasts are supposed to look. More and more, girls at a younger age think they need to dress older than they are, which often means dressing to show off their breasts.

Breasts come in all shapes and sizes, but it's difficult to remember this when you feel like your breasts don't measure up to—or measure far beyond—the other girls around you. Perhaps one breast isn't the same size as the other, or you are flat as a board and think you'll never grow out of your AA bra. Or maybe your breasts are very large, and they are the source of unwanted attention or back pain. Big or small, it's common to be self-conscious about your breasts.

Curvaceous female bodies are normal, but so are slimmer ones. Sophia is a petite, slender girl. Her body, including her breasts, is slower to develop. Read her story to find out how she deals with her experience.

Sophia's Story

Sophia was appalled with her body. While her girl-friends' breasts were starting to grow, she was stuck with a flat chest. She was so flat, she thought her chest was nearly concave. She thought she looked like a boy, and when she felt depressed, she wondered if people mistook her for one. Thank goodness her hair was long.

Sophia had learned recently in health class that family genes have a lot to do with one's body type. Ugh, she thought, and rolled her eyes. Her mother and aunts had small chests. While Sophia still dreamed of exiting puberty with decently sized breasts, she was starting to feel hopeless.

> **While Sophia still dreamed of exiting puberty with decently sized breasts, she was starting to feel hopeless.**

Sophia had never really lacked self-esteem. She was confident and did well in school and extracurricular activities. But she was becoming obsessed with her breast size, which in turn began affecting her self-confidence. She started wearing baggier shirts and her brother's sweatshirts to try to hide what was not underneath.

Talk About It

- Have you started to develop breasts or hips? Are you earlier or later than most of your friends in developing? How does this make you feel?

- Do the women in your family have a certain body type? If so, how do they feel about their body types?

- Do you have a similar body type to the other females in your family? How do you feel about that?

Her mother began to notice, and brought it up one night while they were clearing the dinner table.

"Sophia, honey, why have you been wearing so many of your brother's sweatshirts? Is there a new fashion trend at your school?" her mom asked.

Sophia was annoyed with her mom's intrusion into her style—like she understood anything dealing with fashion anyway.

"No, they're just warmer. Plus, I like how loose and comfy they are."

"You know, you have all those pretty sweaters in your closet."

"I know, Mom. But unfortunately, those pretty sweaters seem to show off that I am hardly a girl!" Sophia exploded. "And you know what? This is all your fault! If you could have given me a different body, everything would be different!" Sophia put the dishes in the sink and stormed off to her room.

Talk About It

- Does realizing that genetics plays a role in your body shape affect what you think about yourself?

- Do you think your body shape will change when you finish puberty? If so, how do you think you'll feel about it?

- Have you ever been so mad or upset about something that you yelled at someone you loved? What did you do?

In between sobs, Sophia heard a knock on her bedroom door. Her mom opened it and poked her head in.

"Soph, can I come in?"

"Whatever," Sophia muttered.

Sophia's mom came in and sat by Sophia on her bed.

"Sophia, I think I understand why you're so upset. When I was your age, I went through the same thing, and I felt like you're feeling now," she said.

"So how did you deal with the fact that you looked like a boy, then?"

"Oh, Sophia, you look nothing like a boy. You are a beautiful girl. I will tell you that I used to stuff my bra around your age. It was ridiculous. My breasts usually looked lopsided and uneven." Sophia cracked a smile at the thought.

"I was a late bloomer in the breast department, like you. But eventually, I did develop. Not too much, of course," Sophia's mom smiled, "but enough to make me feel better about myself."

Sophia's mom told her that she knew Sophia was behind a lot of her girlfriends in developing, but she would catch up eventually. And, until then, the right sort of

"I was a late bloomer in the breast department, like you. But eventually, I did develop. Not too much, of course," Sophia's mom smiled, "but enough to make me feel better about myself."

bra might give Sophia a boost in self-esteem. Sophia nodded, and a smile crept on her face again.

"In fact, I think we're due for a shopping trip," Sophia's mom said. "What do you say to heading to the mall this weekend? We can get you measured and pick out a few special bras for you. How does that sound?"

For the first time in a while, Sophia thought that things were beginning to look a little less bleak.

Talk About It

- Is there someone you can talk to about things that upset you? How often do you talk to this person? Does he or she make you feel better?

- Should Sophia have apologized for getting upset with her mom? Why or why not?

- How do you think a new bra might make Sophia feel about herself?

The underlying shape of our bodies is based on our genetics. While some parts of the body can be altered by exercise and diet, there are other things about our bodies that we just can't change. Chances are likely that your body will resemble another family member's body. It may take time to accept your body and learn that you simply cannot change every part of it.

One great thing that genetics provides for, however, is diversity. No one would want to see only one body type walking around. And although we are told by the media and society that there is one ideal body type, that is simply not the truth. Life would be completely boring if everyone looked the same. However, this idea does not come naturally to many adolescent girls.

Like anything, there are pros and cons to having either a large chest or a small chest. It is learning to embrace the pros and manage the cons that is challenging for girls and for women. For example, one pro for Sophia having a smaller chest is that sports and other activities might be easier and not as cumbersome for her—something she may not have even recognized. Trying to focus on the pros is a healthy start to learning to love your body just the way it is, whatever your situation may be.

Get Healthy

1. Whether your breasts are large or small, the right bra can do wonders for your breasts and self-esteem. Get measured at a store so you can get the right size and fit.

2. Find tops that work with your body type. For example, if you are large-chested, square and small v-neck shirts that are a little fitted but not too tight may be a good option. There are plenty of books and Web sites that can provide tips.

3. If you are being teased in any way about your body, whether it is because of your chest or for another reason, tell a trusted adult. Many schools have programs in place that work to prevent harassment and teasing. It's not okay for anyone to make you feel uncomfortable about your body.

The Last Word from Holly

It seems that breasts can be both a blessing and a curse. They help give females their beautiful curves, but they can also be the source of aggravation and the aim of ridicule. Of course we compare our breasts to others around us and to models we see in magazines. But at the end of the day, what we have is out of our control. So let's be proud of them, no matter how small or large they are.

8

Hair Down There

There comes a time in every girl's life when she looks down, maybe while dressing, and sees a hair or two sprouting beneath her underwear. This is another sign of maturity. Like many other signs of puberty, pubic hair begins to develop around adolescence.

Pubic hair is natural and perfectly normal. Its useful purpose is to protect the body from bacteria. With underwear and clothes, pubic hair may not seem necessary and might seem like a nuisance. There are several attitudes toward pubic hair. Some girls and women find it bothersome and gross, and take measures to get rid of it. Others think of it as a healthy part of their body.

Liz didn't really care either way about pubic hair. To her, it was just something that happened. But when she tried on her swimsuits, she was forced to think about it in a different way.

Liz's Story

"Liz, I am not buying you another swimsuit. You have at least five in your closet that are perfectly fine," Liz's mother said.

"But, Mom, I hate my other ones! They're all old and stupid, and I look ugly in them. I really need that swimsuit with the boy shorts. There is no way I can go to the beach tomorrow in the swimsuits that I have," Liz said.

"That is enough, Liz."

"Mom, come on. I swear this is the only one I'll ask for this year. I won't ask for anything else even. I swear. Please, Mom. I'll do extra chores and promise to keep my room clean."

"Elizabeth, I said

Liz didn't really care either way about pubic hair. To her, it was just something that happened.

no. And you should be keeping your room clean anyway. And since we're on the topic of your room, I can barely see the floor, so go pick it up. Now, I don't want to hear another word about this swimsuit business."

"You don't understand anything!" Liz bellowed as she stormed out of the kitchen and into her room fast enough so she wouldn't get in trouble for talking back to her mother.

Her mom really didn't get a thing, Liz thought. Yes, she had said she needed things before but this time she really, *really*, needed it. Liz was serious. She really did need those boy shorts, or she would not step foot on the beach.

Talk About It

- Do you think Liz handled this situation in a way that will help her get a new swimsuit?

- Think of a time when you wanted something from your parents. How did you go about asking for it? Did it work?

- Do you ever feel like your parents don't understand you? Do you get annoyed, or do you try to help them understand?

After moping around that evening, Liz began rifling through her closet. She whipped out every swimsuit she owned onto her floor and began trying them on. Nothing was working. Again, Liz was getting riled up when the phone rang. It was her best friend, Jessica, calling to talk about the beach.

"Hey, Liz," Jessica said. "I just wanted to make sure we're still going tomorrow. My dad said the weather's going to be awesome. I can't wait. You know, I heard Brad's going to be there. He totally has a crush on you."

Liz thought about Brad. He was so cute. But then she remembered her dilemma.

"Oh, I don't know, Jess," Liz said. "I mean, I really want to go, but I hate all my swimsuits, and my mom won't let me get another one. I mean, they're just, ugh, well they don't fit right. You know?"

"What about the green one you got at the end of last summer? That one's really cute. And you obviously still fit in it. Wear that one," Jessica suggested.

"I can't, Jess. I just can't. You don't understand."

"What?" Jessica asked.

"Well, I can't wear the stupid green suit because I need to wear boy shorts, you know? Like, how do I say this? I can't wear bikini bottoms because, um, well because I have some hair that shows. I didn't realize I'd gotten this much hair until I tried on my suits last week. I don't know what to do, Jess! I can't go to the beach like this! I would just die."

Maybe these things weren't supposed to be talked about. Liz suddenly felt like a complete weirdo.

Jessica was silent on the other end of the phone. Maybe these things weren't supposed to be talked about. Liz suddenly felt like a complete weirdo.

Talk About It

- **Why do you think Liz was nervous to tell Jessica the real reason she was upset? Why do you think Liz hid the reason from her mom?**

- **What do you think Jessica will say?**

"Duh, Liz, you shave it," Jessica answered. "It's not that big of a deal. Just shave that hair like you shave your legs. I've gotta go. We'll pick you up at noon."

After the phone call, Liz thought about what Jessica said for a while. After looking at herself in her swimsuit bottoms one more time, she decided to do it. Liz went into the bathroom, took out the shaving cream and razor from the shower, and shaved her bikini line. After she was done, she looked in the mirror and was quite pleased with the results. The beach would work out after all, she thought.

Talk About It

- **What would you do if you were in Liz's position?**

- **When was the first time you remember noticing your pubic hair? How did you feel about it?**

- **Have you ever shaved your bikini line? Were you happy with the results?**

The next morning Liz woke up in a great mood. She had dreamt about lying on the beach and soaking up the sunshine. She couldn't wait to get to the beach and enjoy the day with her friends.

While still in bed Liz realized her groin area was sort of itchy. She scratched it, but it didn't stop. Liz

kept scratching and then got out of bed and looked at her bikini line area. She gasped when she realized she had red bumps all over her underwear line. What on earth was happening? Liz started freaking out. She then remembered that she had shaved there last night. Oh no! How could she get rid of this before she went to the beach in a few hours?

Liz scrambled into her swimsuit, hoping that the red, itchy bumps wouldn't show—but they still did!

There was no way she could go to the beach looking like this. What could she do?

She had to call Jessica. Jess gave her this stupid idea. She would have to know what to do now. Liz was on the verge of tears when she picked up the phone and dialed the number.

"Jess, I have a huge problem!" Liz exclaimed.

"What is it?" Jessica replied.

"Well, remember how you told me to shave my bikini line last night? I did. And now it's all red and bumpy! There's no way I am going to the beach like this. Tell everyone I got sick, or that I had to babysit my sister. Just don't tell them what happened. It's impossible for me to go looking this way."

Talk About It

- Have you started shaving your legs? If so, have you experienced razor burn? What did you do about it?

- What do you think Jessica will say to Liz? What would you tell Liz?

- Do you think Liz should still go to the beach? What might she learn from this, if anything?

Pubic hair is a way the body protects itself from bacteria and other elements that may enter the body. But in some cultures, pubic hair on women has been looked at in a negative light. Some people associate it with uncleanliness and masculinity. We are told that women should have no hair any place but their heads. That is simply not true. It is there for health and biological reasons.

That said, it's understandable that girls and women would want to stay groomed in order to wear clothes without feeling self-conscious about that area of their bodies. Grooming, trimming, and shaving pubic hair is a personal decision. Whatever you decide to think and do about your pubic hair, make sure you take steps to keep healthy.

Get Healthy

1. If you do decide to groom your pubic hair, take the time to do it in a healthy way. Use clean scissors or a razor. Shave in the direction of hair growth, and take your time. Understand that itching and razor burn can still be an issue.

2. Do not feel like you have to take any actions down below. It's natural to have pubic hair. If you decide to be natural, that is perfectly acceptable.

3. If you are ever questioning whether a part of your body is normal, ask a trusted adult or a doctor next time you have a visit. More than likely, what you're experiencing is another sign of maturity, but hearing this from someone you trust may put your mind at ease.

The Last Word from Holly

At one time or another, most women have probably wished their pubic hair were a little more groomed. But how pubic hair grows differs with each individual. How much hair you have and how dark it is depends a lot on genes, something you can't really change. If you're embarrassed about what you have down there or if you think wearing a bathing suit is the last thing you want to do, for those reasons, maybe you should take steps to groom. But it's important to remember that either choice—staying au natural or grooming a bit—is perfectly healthy and normal.

9

First Period

One of the biggest changes adolescent girls experience is getting their periods. Menstruation is regulated by hormones, and many girls get their period for the first time between the ages of 10 and 16. Some girls get their first period earlier, and some get it later. Some women's periods last only three days, and some last up to seven. There is so much variance in these numbers because every woman is unique!

A common concern for adolescent girls is where or when they'll get their first period. We have all heard horror stories about a girl getting her period while in class and leaking through her white pants for everyone to see. However, this rarely happens. First, most girls do not have a very heavy flow for their first period, which means there wouldn't be enough to leak through your pants. Second, you will probably feel wetness in your underwear, and you'll be able to take care of the situation before something embarrassing happens.

Katie's story is very common. Getting her first period at school was more of an exciting event to share with her friends than a bad dream.

Katie's Story

"Does anyone know the answer to the algebra problem on the board?" asked Mr. Gullingsrud. "You've had some time to work on it. It's time to discuss some amazing math solutions!"

Katie groaned quietly in her seat. She couldn't handle Mr. Gullingsrud's pep today. Her stomach had been aching for hours. She thought it was hunger pains, even though she'd eaten a good breakfast of cereal and orange juice that morning. Thankfully there were only a few more minutes until the lunch hour. Katie pushed her hands into her stomach and leaned over a

Katie groaned quietly in her seat. She couldn't handle Mr. Gullingsrud's pep today.

little to relieve the aching. She pretended to be listening to Mr. Gullingsrud until the bell rang, and then breathed a sigh of relief as she left the classroom.

Talk About It

- Can you think of a time when you didn't feel well and had to be in class? What was wrong? Was it hard to stay focused?

- What do you think is causing Katie's stomach pains if she ate a good breakfast?

Kaia, Nicole, and Tyanne were waiting for Katie at her locker. The four usually walked to lunch together because their classes were in the same area of the school.

"Hi, Katie," Tyanne said. "How was algebra?" Katie gave a grimaced smile and mumbled an okay.

"Hey, are you feeling all right? You look a little strange," Kaia said as she examined Katie, who was hunched over a bit.

"Yeah, I think I'm okay. My stomach hurts a little bit. I think I'm really hungry or something. I don't know. Let's just go to the cafeteria," Katie said.

"Sounds good to me. I'm starving!" Nicole responded. "But first can we stop at the bathroom on the way? I really have to pee."

The girls laughed at Nicole's bluntness and all agreed that they could make the stop.

Talk About It

- Do you have a group of friends who you have a lunch or other routine with at school? What do you do?

- Do you think Katie's friends should be concerned about Katie's stomach pain?

- Have you ever had to take care of a sick friend?

When the girls reached the bathroom, Nicole suggested that Katie go as well.

"What? Why?" Katie asked, confused. She didn't think she'd mentioned having to go before.

"Well," Nicole said, with a grin spreading across her face, "sometimes when I have a stomachache, I have to go to the bathroom. You know," she laughed.

The three girls smiled and shook their heads. But Katie realized Nicole had a point. Plus, she did have to pee a little. She didn't notice until now because she'd been too focused on her stomach.

But when Katie was in the bathroom stall and had pulled down her pants and underwear, she saw a few dark red spots on her underwear. She was alarmed for a second, but then she realized what it was.

"Oh, jeez," Katie sighed, loud enough for Nicole to hear her in the next stall. Nicole asked Katie if everything was okay.

"Well, I think I just got my period."

Talk About It

- How do you think Katie felt when she saw the dark red spots on her underwear? What do you think went through her mind?

- How do you think Nicole will react? What about Katie's other friends?

"You what? All right!" Nicole had flushed her toilet and was out of the bathroom in seconds. "Do you have a pad or anything? Because I have one in my locker. Hold on, I'll go get it." Nicole raced out of the bathroom.

While Katie waited for Nicole to come back, she remembered the talk she had with her mom about this. Her mom told her that this would be an exciting time and was not something to get nervous or scared about. The best thing to do was to be prepared, and Katie was.

She had a pad in her bag for this situation. But Nicole was so excited, she ran out of the bathroom before Katie could tell her she had a pad in her backpack. Oh well, Katie thought. Nicole did seem pretty eager to help. And Katie could use the pad in her bag later, or when one of her friends needed it next.

When Katie and Nicole walked out of the bathroom, Kaia and Tyanne were waiting for them.

"So," Kaia asked, hesitantly, "are you guys ready to go to lunch?"

Talk About It

- If you've already gotten your period, how did you first notice it? Was it exciting, embarrassing, or nerve-racking?

- If you haven't had your first period, are you prepared for it? How? Have you talked about this time with your mom or another adult you trust?

- Have you ever had to help a friend in this kind of situation? What did you do?

Katie knew that Kaia and Tyanne already guessed what happened. She decided not to keep it a secret. After all, they were her good friends, and they would all be going through the same thing soon—if they hadn't already.

"Well, this is sort of embarrassing to say, I guess, but I, um, I got my period," Katie said. "So, I have my period now. So that's why you saw Nicole running like a maniac to her locker and back to the bathroom. Nothing's wrong and all."

There was a short pause as Katie looked at her friends. Then Tyanne spoke up.

"That's cool, Katie. I'm kind of excited, but my mom can't stop talking about it. She's more excited about it than I am."

"I haven't gotten mine yet either, but I don't want to," said Kaia. "What a pain."

"It's really not that bad," Nicole said. "I've had my period for two years now. I got it when I was ten. Yeah," Nicole said, addressing the surprised looks of her friends, "I got mine super young, I guess. And at first it was weird to get used to, but after a little bit it was just a normal thing. I'm so excited you got yours, Katie. Now I have someone to talk to!"

Katie felt her stomach rumbling and was truly hungry now. She suggested they go to the cafeteria and eat before lunchtime was over.

While the girls walked, Katie thought about telling her mom that night. She knew her mom would be excited to hear the news. No doubt they would take a trip to the store to get some "products," as her mom called them.

She was thankful for her friends' support. She'd heard some awful stories of girls getting their periods for the first time. Katie was surprised to realize that she wasn't at all embarrassed about the events of the day. It was sort of a funny experience, actually, with Nicole running around to help her out. Katie was glad that her first period was fairly uneventful, because a dramatic horror story was definitely something she could do without.

Talk About It

- Have you talked with your friends about periods? Do you know if any of your friends have gotten theirs?

- If you were around when one of your friends got her period for the first time, what would you do and say?

- If you have gotten your period, do you have any symptoms that help you know when it's coming? If so, what are the symptoms?

Ask Dr. Robyn

For ages, menstruation has been something discussed behind closed doors, or not at all. There have been negative attitudes toward it, with some people thinking it's unclean or a cause for embarrassment or shame. But times have changed, thank goodness, and it's now more often seen as a wonderful development girls go through as they grow older.

Some girls may experience physical symptoms before they get their period each month. These symptoms may include headaches, bloating, or cramps. If you feel discomfort before or during your period, there are a number of things you can do to feel better. Ask a parent to help you find a medicine that is made specifically to reduce these symptoms. Some girls also find that heating pads can help relieve their cramps.

The beginning of this time—the first period—can be a big moment in a girl's life. It can also happen at very different ages. It's important to remember that your body knows when it's ready, and not to freak out if you're the early or late bloomer of your group of friends. It will happen. Don't worry. And after it does happen, don't be afraid to talk about it. It is an exciting thing, and talking about your shared experiences with other women might help all these changes seem a little less scary.

Get Healthy

1. Be prepared. Keep a couple of pads or tampons in your locker or purse so you are ready when your first period comes.

2. If you're nervous about getting your first period, talk with a trusted adult or friend who may be able to shed some light on the event.

3. If you're a late or early developer, trust that your body knows when it's ready for its first period. Although you may feel uncomfortable getting your period at an early or late age, your body does know best.

The Last Word from Holly

The way I know my first period was not a life-altering catastrophe is because I don't remember the experience at all. If it had been embarrassing, I would have some memory of it, but I don't. It's sort of odd, I know, but I think it must have been similar to Katie's experience. I got my period and was prepared for it and that was that. You see, more proof that those embarrassing urban legends are exactly that: urban legends.

A Second Look

From the stories in this book, I hope you will begin to realize that what you and your body are experiencing are changes that happen to all girls. During adolescence, the most common thing you will experience is going to be change. And, unfortunately, it won't always be easy or comfortable, especially when so much of it comes at once.

There are outward changes that others can see. Your body shape will begin to develop in more characteristically female ways, you may get a bad haircut or braces, and you may start to get acne. But there are also less-visible changes that you might think you're going through alone. Hair will grow in places that didn't have hair before, you'll begin menstruating, you might start to sweat more often and possibly smell a bit. It is all a lot to handle, but it may seem even worse if you think you're going through it alone. You're not. And if you don't talk to others about the changes you're experiencing, I hope you can at least look to the stories here to know that the changes are all normal.

Thinking about your body and the new things you're facing may take up a lot of your time. But try to find other things to think about and do as well. Make time for your family and friends. Spending time with others, especially other girls who are going through similar things, can be a great release of tension and confusion. Hanging out with friends can put your focus on fun. It may also help form a kind of support group for you to turn to when you are overwhelmed by changes. Remember, you don't have to go through this alone.

XOXO,
Holly

Pay It Forward

Remember, a healthful life is about balance. Now that you know how to walk that path, pay it forward to a friend or even to yourself! Remember the Get Healthy tips throughout this book, and then take these steps to get healthy and get going.

- Be proud of who you are as a person. Take time to develop your unique self, and your appearance will become secondary. Your body is just a shell. Don't make it into all that you are.

- Try not to let media images of girls and women affect how you think about your body and self. Remember that magazines and media are heavily retouched and what you see there is not real life.

- If you have questions or doubts about anything that is going on with your body, talk to a trusted adult or a doctor about your concerns.

- Adolescent body changes fall in a large age range. You may be an early or late developer, but being either does not make you abnormal. Your body knows best when it is ready for changes.

- Practice good hygiene. Floss and brush your teeth, and wash your face, hair, and body regularly. Doing these things will decrease plaque, acne, and body odor, and will leave you feeling fresh and clean and ready to take on the world.

- If you are uncomfortable with your skin imperfections, try using makeup to cover them up or make them less visible. But don't get hung up on a breakout or a birthmark. Stressing or picking at your skin will only make it worse.

- Wear clothes, including bras, that fit well, feel good, and flatter your body. What works well on one body type may not work at all on the next. Just because something does not fit you in the right areas does not mean you are too fat or skinny or have a weird body. All bodies are different, so different clothes work on different girls.

- If you are being teased, picked on, or bullied for any reason, including for how you look, tell a teacher or another adult. On the flip side, don't bully others for any reason, and don't join in on picking on an "easy target." Nobody likes to be teased, so don't encourage it or participate in it.

Additional Resources

Select Bibliography

Madaras, Lynda, and Area Madaras. *The "What's Happening to My Body?" Book for Girls*. New York: New Market Press, 2007.

McCoy, Kathy, and Charles Wibbelsman. *The Teenage Body Book*. New York: Hatherleigh, 2008.

Redd, Nancy Amanda. *Body Drama*. New York: Gotham, 2008.

Further Reading

Kirberger, Kimberly. *No Body's Perfect: Stories by Teens about Body Image, Self-Acceptance, and the Search for Identity*. New York: Scholastic, 2003.

Le Jeune, Veronique, and Philippe Eliakim, with Melissa Daly. *Feeling Freakish? How to Be Comfortable in Your Own Skin*. New York: Amulet, 2004.

Madaras, Lynda, and Area Madaras. *My Body, My Self for Girls*. New York: New Market Press, 2007.

Zimmerman Rutledge, Jill S. *Picture Perfect: What You Need to Feel Better about Your Body*. Deerfield Beach, FL: Health Communications, Inc., 2007.

Web Sites

To learn more about dealing with physical changes, visit ABDO Publishing Company online at **www.abdopublishing.com**. Web sites about dealing with physical changes are featured on our Book Links page. These links are routinely monitored and updated to provide the most current information available.

For More Information

For more information on this subject, contact or visit the following organizations.

The Dressing Room Emerging Women Projects

45 State St., #280, Montpelier, VT 05602
828-318-4438
www.thedressingroomproject.org
This initiative's goal is to let girls know that they should not judge themselves based on media's standards of beauty. Included on the site are information on workshops, resources, and ways to take action.

New Moon Girls

2 West First Street, #101, Duluth, MN 55802
800-381-4743 or 218-728-5507
www.newmoon.com
This online site and magazine shows girls that they can express themselves in an individual and meaningful way. The site displays writings and artwork done by adolescent girls and has a feature on body and feelings.

Glossary

birthmark
> A mark on the skin, usually present from birth, caused by a cluster of pigmented skin cells.

concave
> Hollow and curved inward.

dermatologist
> A type of doctor who works specifically with skin and skin conditions.

dilemma
> A serious problem.

empathize
> To share another's emotions, thoughts, or feelings.

equilibrium
> Balance.

genetics
> A person's hereditary makeup and inherited characteristics.

lethargic
> Having decreased energy; sluggishness.

media

Any of the methods of mass communication, including television, radio, newspapers, books, and magazines.

menstruation

The bleeding that women experience for approximately three to seven days every month; also called a period.

orthodontist

A type of dentist who works with braces.

pigmentation

Coloring of the skin.

retouch

To alter a photograph so that the imperfections are hidden.

urban legend

A story that is passed from person to person and is often regarded as true, even if it is not.

Index

About the Author

Holly Saari is an editor who works with children's educational books in Minnesota. She is also very interested in helping girls improve their body image and self-esteem. This is her first book.

Photo Credits

iStock Photo, 12, 54, 94; Eileen Hart/iStock Photo, 14, 52; Quavondo Nguyen/iStock; Photo, 17; Ron Sumners/iStock Photo, 22; Ilya Rabkin/Shutterstock Images, 24; Phil Date/iStock Photo, 27; Jennifer Trenchard/iStock Photo, 33; Hannamariah/Shutterstock Images, 35; Adam Goodwin/iStock Photo, 37; Brian McEntire/iStock Photo, 42; Catherine Lane/iStock Photo, 45; Eva Serrabassa/iStock Photo, 47; Ana Abejon/iStock Photo, 57, 73; Alberto Pomares/iStock Photo, 63; Jim Jurica/iStock Photo, 65; Roxana Gonzalez/iStock Photo, 71; Shutterstock Images, 75; Liv Friis-Larsen/iStock Photo, 81; Monkey Business Images/Shutterstock Images, 83; Graca Victoria/Shutterstock Images, 85; Bill Lawson/Shutterstock Images, 87; Galina Barskaya/Fotolia, 92; Chris Schmidt/iStock Photo, 99